Designing for All Life

A Manifesto for Regenerative Design Across Species, Systems, and Time

His Majesty King Duncan

Designing for All Life

His Majesty King Duncan © 1995

Table of Contents

Preface — A World Designed for All Life

I dedicate this book to my parents Her Majesty Queen Lizzy, His Royal Highness Prince John Charles Wright, to my wife Her Majesty Queen Agnes, my children and family.

"The future is not designed by those who seek to control it, but by those who learn to care for it."

This book began as a question whispered in the midst of a planetary   crisis:
What if everything we designed — every object, city, policy, and digital system — were designed not just for human comfort, but for the continuation of life itself?

For centuries, design has served the human project: our convenience, our economy, our aesthetics. But that narrow focus has also brought us here — to melting glaciers, vanishing species, rising inequality, and a planet trembling under the weight of our own inventions.

It   is   no   longer   enough   to   design   for   users.
We must design for **life**.

This book is a call — and a map — for a new kind of thinking: **expansive   design**.
Design that stretches beyond the immediate and the individual. Design that listens to the land, learns from the wind, collaborates with   rivers,   and   makes   kin   with   machines. Design that expands not just in form or function, but in empathy.

Over the years, I have watched designers, ecologists, and communities from Scotland to Zimbabwe, from the Swiss Alps to the deserts of Dubai, experimenting with this broader design consciousness. They are building schools that feed ecosystems, houses that heal, and policies that protect the unborn. They remind us that design is not a profession — it is a practice of relationship.

This is not a manifesto of perfection. It is a meditation on possibility — on how our creative choices can once again align with the living systems that make existence possible. It is an invitation to slow down, to reimagine success, and to design with humility, awe, and continuity.

Because one truth has become undeniable: If our designs do not sustain life, they will eventually design life out of existence.

Let us design as if the world were alive — because it is.

—   *HM   King   Duncan*
London / Edinburgh / Zurich / Durban / Mutare / Dubai

Chapter 1 — Time: Designing Beyond the Present

"The future is not a destination; it's a responsibility." — Unknown

The Long View of Design

Most design today operates at the speed of markets: quarterly reports, product cycles, election terms. But the consequences of design—plastic in the ocean, emissions in the sky, and algorithms in our culture—persist for generations. If we are to design for all life, we must learn to design **with time**, not against it.

To think expansively about time means stretching our imagination beyond the next launch or fiscal year, into centuries and cycles. It means seeing design as an act of **care across generations**, a way of leaving traces that nourish rather than deplete. Time, in this sense, becomes a design material—fluid, living, and deeply moral.

Designing in Deep Time

In Cornwall, UK, the **Eden Project** sits inside a reclaimed clay pit—a wound in the Earth reborn as a sanctuary for biodiversity and education. Its vast biomes are living time capsules, each containing ecosystems from distant eras and climates. The project didn't just rehabilitate a site; it reimagined a legacy. Eden invites visitors to see time layered—industrial, ecological, human—and to imagine futures of renewal.

In **Scotland**, the *Orkney Islands* lead the world in marine renewables. Here, time flows with the tides. Projects like the

European Marine Energy Centre test devices that will outlast oil rigs and redefine coastal economies for a century to come. The design of these systems isn't about immediacy; it's about **inheritance**—what energy, culture, and ecosystems we bequeath to the future.

And in **Switzerland**, the *Monte Rosa Hut* clings to a glacier above Zermatt, glimmering like a silver shard of time. Built by ETH Zurich, it's nearly self-sufficient: solar-powered, rain-harvesting, and designed for 100 years of mountain change. The architects embraced **climate adaptation** as a design principle. When the glacier melts, the hut remains—an artifact of resilience and foresight.

Urban Time and Circular Futures

Across **Europe**, cities are learning to design for long-term circularity rather than short-term consumption. Amsterdam's *Circular Economy Plan 2050* is not just policy—it's a design philosophy. Every new development must account for material reuse, carbon cycles, and social equity. Time, here, is measured not in profit but in **ecological duration**: how long something contributes positively before it decays.

Dubai's Sustainable City, meanwhile, demonstrates temporal design in an opposite context—a fast-moving desert metropolis. Its developers asked a radical question: can a city built in haste still serve generations? The answer: yes, if every layer—solar, water, food, mobility—is designed to renew itself. The city's closed-loop systems show that speed and sustainability need not be enemies; when aligned, they create **regenerative momentum**.

Indigenous and Ecological Time

In **Zimbabwe**, time has always been cyclical. The land operates on ancestral rhythms of rain, soil, and spirit. The *Kariba REDD+ Project*, one of the largest forest conservation programs in Africa, protects nearly 800,000 hectares of woodland from deforestation. Its design is not architectural but ecological and social—aligning community development with carbon storage. Villagers plant trees they may never sit beneath; their children inherit both the shade and the stewardship.

This way of designing—*for those not yet born*—is ancient wisdom returning to modern practice. It reminds us that sustainability is not innovation but remembrance.

Designing for the Next Century

Time also tests the durability of ideas. In the age of planned obsolescence, longevity is rebellion. Swiss watchmakers have long embodied this philosophy: craftsmanship as continuity. But today, that ethic extends to software, infrastructure, and governance.

In *Zurich*, urban designers employ "100-Year Plans" to guide city development, ensuring that housing, transit, and green spaces evolve gracefully rather than abruptly. These aren't static blueprints but living frameworks that adapt like ecosystems. Time becomes a co-designer—a partner that refines, erodes, and renews.

In *Scotland's peatlands*, restoration projects mirror this patience. Peat forms at a rate of one millimeter per year, meaning the moss we see today began growing when Shakespeare wrote his plays.

Restoring these landscapes demands humility: the designer must think in millennia, not months.

Temporal Ethics

What if design ethics included the unborn? What if every blueprint carried a **time signature**: a record of who will benefit—and who will bear the cost—decades from now?

Some European designers now include "legacy audits" in their processes, evaluating how a product or building will age, degrade, or return to the biosphere. Others advocate for **temporal reparations**—restoring the damage inflicted by past industrial design through future-positive action.

In this expanded time horizon, the designer becomes a **steward of duration**—one who tends not only to form and function, but to the invisible thread between generations.

A Practice of Patience

In a Zurich laboratory, engineers test new materials that biodegrade within precise timeframes—months for packaging, years for construction, centuries for archives. In Harare, young designers use recycled steel from colonial-era railways to craft modern furniture—transforming the past into living design.
In the Scottish Highlands, architects build with timber that will naturally rejoin the soil, leaving no ruin behind.

Each is a gesture of respect toward time: nothing is permanent, but everything matters.

The Temporal Imagination

Time-centered design asks us to **slow down**, but also to **stretch forward**. It asks questions that unsettle comfort:

- How will this place breathe when I am gone?
- Will this system nourish or deplete?
- Can my design be a good ancestor?

The answer lies not in technological foresight alone, but in moral foresight—the courage to imagine lives beyond our own.

Closing Reflection

In a world obsessed with the now, **designing for time** is radical compassion.
It is the choice to create not for the next user, but for the next century; not for the market, but for the meadow.

Every building, algorithm, and policy we design becomes a message to the future.
Let it say: *We cared enough to think ahead.*

Chapter 2 — Proximity: Rethinking Closeness in a Connected World

"We are more connected than ever, yet more divided than we can bear." — Anonymous

The Paradox of Closeness

Proximity once meant nearness — bodies, voices, streets, shared breath. Today, connection has become weightless: a click, a screen, a signal. We live in a paradox of intimacy and isolation.

To design for all life, we must learn to **redesign proximity** — to restore depth to our connections while expanding our circles of empathy beyond the human. The question is no longer *how connected we are*, but *how meaningful those connections are.*

Design becomes an instrument of **belonging** — building bridges across distances of geography, culture, species, and emotion.

Local Connection, Global Meaning

In **London**, the *Library of Things* is a modest revolution. Instead of each person owning drills, sewing machines, or tents, neighbors share them. Objects move between homes; so do conversations, stories, and advice. It's not just about sustainability — it's about **social fabric**. Design here repairs both material and relational waste.

In **Scotland**, small communities in the Highlands and Islands are experimenting with **rural co-working hubs**. These spaces fuse digital technology with local heritage, allowing islanders to work globally without abandoning their roots. The architecture is simple — wooden walls, natural light, shared kitchens — but the design is profound: *belonging without borders.*

Meanwhile, in **Zurich**, *civic co-design labs* invite citizens to shape urban policy through participatory workshops. Proximity becomes democratic — not physical distance, but the closeness between people and power. When residents can co-design their futures, cities become shared acts of authorship.

Proximity in Public Space

In **Barcelona**, the *Superblocks* initiative reclaims city streets from cars and returns them to people. What was once traffic is now playground. Noise becomes laughter. Air becomes breathable. The design is infrastructural, but the outcome is emotional: neighborhoods rediscover the pleasure of being near.

Harare, Zimbabwe, is witnessing its own form of spatial closeness. In community-driven design workshops, citizens collaborate with young architects to co-create markets, bus shelters, and gardens.
These aren't just public spaces — they're **spaces of participation**, where design is less about building and more about belonging.

And in **Dubai**, where distances are vast and temperatures extreme, the *Museum of the Future* and *Al Seef Waterfront* create emotional nearness through **immersive design**. Visitors move through multisensory narratives — light, scent, sound — that dissolve cultural and generational gaps. The future feels close enough to touch.

Designing Emotional Proximity

Closeness isn't only physical — it's emotional, even spiritual. During the pandemic, distance became protection. Yet the loneliness that followed revealed how deeply we rely on presence. Designers began rethinking empathy — how to feel near without being near.

In **Scotland**, *Glasgow's community arts programs* turned design into healing. Through participatory murals, neighborhood photography, and co-created exhibitions, residents reconnected after isolation. Art became architecture for emotion.

In **Switzerland**, *care design* is becoming an emerging field. Hospitals and clinics are being reimagined as **spaces of dignity**, where patients feel seen rather than processed. Proximity, here, is measured in attentiveness — a nurse's eye contact, the warmth of materials, the softness of light.

The Digital Dilemma

Digital networks have made proximity infinite — but also superficial. A million connections can equal zero community. Expansive design thinking calls for **"slow technology"** — tools that invite reflection rather than reaction.

In **Europe**, projects like *Time Well Spent* and *Digital Humanism Vienna* promote humane interfaces that value depth of interaction.

Interfaces are redesigned to encourage pause, dialogue, and empathy.

In **Dubai**, smart city systems use AI to make life efficient — but the challenge is keeping it **humane**. Designers now embed ethical guidelines ensuring that algorithmic decisions preserve trust and inclusion. Closeness must never be traded for control.

And in **Zimbabwe**, mobile networks serve as platforms for local storytelling. Through apps like *ZimboHub*, young creatives share poems, crafts, and films, reconnecting rural and urban communities through shared imagination — **digital proximity as cultural renewal**.

Nature as Neighbour

What if proximity extended beyond people? If design is to serve *all life*, closeness must include the more-than-human world: soil, trees, rivers, insects.

In **Switzerland's Sihlwald Forest**, visitors experience a new kind of park — one left mostly untouched. Paths are designed to bring humans close without intrusion. The design teaches respect: the less we interfere, the closer we truly become.

In **Scotland's rewilding zones**, hikers are invited to walk "quiet trails," designed with minimal signage, where the goal is not navigation but **attunement** — listening, noticing, coexisting. This is proximity as empathy — the closeness that comes from stillness.

Global Interdependence

Proximity also means recognizing **interdependence** across nations and ecosystems. A drought in Zimbabwe affects grain prices in Europe. An innovation in Swiss renewable energy influences desert housing in Dubai. A policy in Scotland can inspire global biodiversity strategies.

Design must reflect this **planetary proximity** — a world where every local act ripples through global systems. The illusion of distance has vanished; the future is shared.

Designing for Intimacy

How, then, do we design for closeness in a fragmented age? The answer lies in **invitation** — creating systems that bring people, ideas, and species into relationship. A co-housing community, a circular marketplace, a rewilded park, or even a humble tool library — all are acts of design that **shorten the emotional distance** between beings.

The designer becomes a facilitator of nearness — someone who crafts the conditions for contact, care, and co-existence.

Closing Reflection

We are living in an era of invisible distance. Screens connect but separate; cities grow but isolate. Yet design offers a way home — not backward, but inward. To design for proximity is to **design for empathy** — for the touch that mends, the dialogue that deepens, the neighborhood that breathes together.

Every bridge, interface, and street we design whispers a question: *How close do we dare to be?*

Chapter 3 — Value: Redefining What Matters

"We measure what we treasure — and too often, we treasure what we can measure." — Kate Raworth

The Currency of Design

Every design decision expresses a value. It reveals what we believe is worth protecting, producing, and perpetuating. For too long, the value system that has guided design has been narrow — economic growth, efficiency, and speed. The results have been astonishing but costly: climate collapse, inequality, and alienation.

Expansive design asks a new question: **What if value were measured in life, not in money?**

To design for all life, we must **redefine value** — shifting from extraction to regeneration, from profit to purpose, from ownership to stewardship.

The End of Extractive Design

The 20th century was the age of extraction: from oil wells to data mines, the world's wealth was built on removing more than we returned. Designers became skilled at making the unsustainable look irresistible.

But the 21st century demands a new design language — one that values **restoration, reciprocity, and responsibility**.

In **London**, the rise of the *B Corp movement* marks this transition. Certified B Corporations commit to measuring success not by profit, but by social and ecological impact. From design agencies to consumer brands, companies are rewriting what it means to win. Value becomes a moral ecosystem, not a spreadsheet.

In **Scotland**, *Social Bite cafés* embody this ethos at street level. Founded to end homelessness, they operate as social enterprises, reinvesting profits into jobs and housing. Each sandwich sold is a small act of redistribution — value flowing back into community.

Value as Relationship

True value is relational. It emerges not from what we own, but from **how we relate** — to people, place, and planet.

In **Switzerland**, a country synonymous with precision and quality, this idea takes tangible form. The *Swiss Made Green label* expands the meaning of national craftsmanship to include sustainability. Products carrying the label must meet strict environmental standards — proof that beauty and integrity are inseparable. Here, value is measured in care.

In **Zimbabwe**, this relational perspective has deep roots. Traditional Shona philosophy speaks of *ukama* — the idea that "to be is to be related." Design inspired by *ukama* treats value as a web of mutual benefit. When communities build solar kiosks or irrigation systems, as seen in *Econet's Solar Project*, the aim is not ownership but continuity — ensuring that power, both literal and social, circulates.

Beyond GDP: The Value of Well-Being

Across **Europe**, a quiet revolution is underway. Economists, architects, and policymakers are rejecting GDP as a measure of progress and embracing **well-being indices** instead.

In *Amsterdam's Doughnut Economy*, urban design aligns with ecological and social thresholds. Projects must meet human needs without breaching planetary limits. The "sweet spot" — the doughnut's ring — represents equilibrium: prosperity within boundaries.
This is design as balance, not excess.

Scotland became the first country to join the *Wellbeing Economy Alliance*, embedding compassion into national policy. Government initiatives now assess how decisions affect mental health, community cohesion, and nature. In this framework, the designer becomes part economist, part healer — shaping not just spaces, but societies.

The Aesthetics of Enough

Luxury once meant excess — marble, gold, spectacle. But in a world of scarcity, luxury is being redefined as **enoughness**. Designers are rediscovering the beauty of restraint, durability, and repair.

In **Switzerland**, architects like Peter Zumthor craft buildings that age gracefully. His *Therme Vals* is built from local stone, drawn from the mountain itself. Over time, it absorbs the mineral hues of the landscape. Value, here, is temporal — it deepens with age rather than fading.

In **Dubai**, sustainability is becoming the new prestige. The *Alserkal Avenue* arts district, housed in repurposed warehouses, celebrates reuse as refinement. The design demonstrates that cultural value can emerge from **honesty of materials** and **continuity of purpose** — a quiet rebellion against disposable opulence.

And in **Zimbabwe**, artisans weaving baskets from reclaimed wire or sculpting from serpentine stone demonstrate that value is born from **ingenuity**, not abundance. Craft becomes economy, art becomes livelihood.

Ethical Economies

In the new value landscape, transparency and accountability are design   features. Consumers want to know the story behind what they buy: Who made it? Under what conditions? With what impact?

Swiss watchmakers, long associated with precision, are now embedding blockchain systems to trace sourcing of metals and ethical labor. The invisible becomes visible — value becomes verifiable.

In **Europe**, design firms are creating "impact manifests" alongside their portfolios, listing carbon footprints and social metrics for each project. The creative process becomes an ethical conversation.

In **Dubai**, architecture firms now compete on environmental certification as much as aesthetics. Sustainable design isn't just moral — it's marketable. Value has evolved into virtue.

Sacred and Shared Value

Perhaps the most profound shift in value is **spiritual** — recognizing that not everything valuable can be priced. Clean air, fertile soil, coral reefs, cultural memory — these are the invisible currencies that sustain all life.

In **Scotland**, community land trusts allow residents to collectively own and manage local terrain. Land becomes commons again — a shared inheritance, not a commodity. In **Zimbabwe**, sacred groves are protected not by law, but by reverence. They remind us that value is sometimes sacred precisely because it is *untouchable*.

In **Switzerland's Lavaux Vineyards**, a UNESCO World Heritage site, stewardship is ritual. Families have cultivated the same terraces for centuries, guided by seasonal rhythms. Each stone wall holds both labor and love. The design is agricultural, but the outcome is spiritual: continuity as culture.

Designing the Future of Value

What might a *design economy for all life* look like?

It would reward restoration over extraction. It would prize empathy over efficiency. It would see the invisible as indispensable — from microbial life in soil to the emotional life of a city.

Imagine:

- Buildings that **generate value** for ecosystems by purifying air and water.

- Digital platforms that measure **collective happiness** instead of engagement metrics.
- Financial systems that **circulate gratitude** alongside currency.

This is not utopian — it's necessary. The collapse of old value systems opens the door to new forms of prosperity.

Closing Reflection

The designer of the future is an **alchemist of value** — transforming waste into worth, isolation into connection, consumption into contribution. To design for all life is to understand that value is not something we extract from the world, but something we return to it.

The most valuable thing design can create now is **meaning** — meaning that sustains, includes, and endures.

Chapter 4 — Life : Designing for Biodiversity, Ecosystems, and Non-Human Stakeholders

"When we try to pick out anything by itself, we find it hitched to everything else in the universe." — John Muir

The Center of Everything

To design for all life is to design for everything that breathes, grows, crawls, and decays. This chapter sits at the heart of *Designing for All Life*, because *life* itself is the only true client of the future.

For centuries, design has been proudly *human-centered*. We made the world legible, comfortable, and efficient for ourselves — often at the expense of other species and systems. Cities sprawled over wetlands. Materials were mined faster than they could regenerate. Rivers were diverted, forests cleared, oceans choked.
In our pursuit of convenience, we forgot that our comfort depends on everything else being alive.

But now, a shift is emerging — from *human-centered design* to **life-centered design**.
This is not a rejection of humanity, but a recognition that we are part of a far larger design system — the biosphere itself.

1. From Human-Centered to Life-Centered

Traditional design asks: *What do people want?* Expansive design asks: *What does life need?*

Life-centered design redefines the very brief of creation. It recognizes that every product, structure, and system participates in the metabolism of the planet. Nothing we make is neutral; everything has ecological consequence.

In **Scotland's Cairngorms National Park**, the *Cairngorms Connect* initiative exemplifies this philosophy. It's the UK's largest rewilding project — 600 square kilometers of forest, moor, and wetland restored to their natural rhythms. Designers, ecologists, and communities collaborate not to impose form but to **listen to the land**. Fallen trees are left to rot. Rivers are allowed to meander again. Deer populations are balanced with wolves and lynx.
Design becomes less about shaping and more about *allowing* — the art of stepping back so life can step forward.

Similarly, in **Switzerland**, the *Sihlwald Wilderness Park* near Zurich has been left untouched for over a decade. Trails and signage are carefully designed to minimize disturbance. Instead of manicuring the forest, managers design **for decay** — recognizing rot as renewal, and death as design. Visitors are invited to see decomposition as creation.

Life-centered design, then, is not merely aesthetic. It's ethical, ecological, and existential.

2. Designing for Biodiversity

Biodiversity is not a luxury — it is the planet's operating system. Every species, visible or invisible, performs a function in the web of life: pollination, filtration, decomposition, renewal. When we design for biodiversity, we design for **resilience** — systems that can adapt, evolve, and survive disruption.

In **the UK**, the *Wild Ennerdale* project reimagines what a national park can be. Instead of managing landscapes for human enjoyment, it restores ecological autonomy. Engineers removed riverbanks to let water carve its own course. Over time, salmon returned, trees spread, and flood risk decreased naturally. The best design decision was restraint.

In **Berlin**, urban planners have created *wildlife corridors* threading through dense neighborhoods, allowing foxes, hedgehogs, and birds to move freely. Architecture here becomes **habitat**, not boundary. Green roofs, insect hotels, and rain gardens transform cities into multispecies homes.

In **Dubai**, the *Ras Al Khor Wildlife Sanctuary* demonstrates life-centered design in an extreme environment. Amid skyscrapers and highways, mangroves flourish. Designers built observation hides shaped like traditional barjeel wind towers, letting people witness thousands of flamingos without intrusion. The sanctuary filters urban runoff, cleans air, and cools the city — a reminder that even in hyper-modern landscapes, life finds a place if given space.

And in **Zimbabwe's Vumba Mountains**, communities and ecologists collaborate on forest restoration through the *Vumba Restoration Initiative*. Instead of monocultures, they replant native species, restoring bird habitats and soil fertility. Local carpenters craft tools from fallen timber only, ensuring the cycle remains regenerative. Here, design and ecology speak one language: renewal.

3. Ecosystems as Co-Designers

What if rivers, fungi, or coral reefs were recognized as **collaborators** in design — not resources, but participants?

Across **Europe**, a movement called *Bioregional Design* is taking hold. Cities like Amsterdam, Milan, and Geneva are mapping themselves not by administrative boundaries but by **watersheds, soils, and ecosystems**. Urban plans are written in the grammar of geology and water flow. Design begins where nature dictates.

In **Switzerland**, ETH Zurich's *Living Materials Lab* explores biodesign — growing construction materials like mycelium bricks and algae facades that sequester carbon. Buildings become **photosynthetic partners**, not inert objects.

Meanwhile, **Zimbabwean permaculture cooperatives** treat farms as dynamic systems, where chickens, compost, and crops all co-design fertility. Waste becomes input. Diversity becomes strategy. This is biomimicry in practice — **design that behaves like life**.

4. Non-Human Stakeholders

Every design has stakeholders who cannot speak: rivers, trees, insects, animals, and the unborn. Life-centered design gives them **representation**.

In **New Zealand**, the Whanganui River was granted legal personhood — it can sue and be defended in court. Similar movements are rising across Europe and Africa. Imagine extending that logic globally: rivers, forests, and glaciers treated not as property, but as kin.

In **Switzerland**, debates around the *Rights of Nature Initiative* reflect this shift. If adopted, ecosystems could gain constitutional recognition — a radical reimagining of who "belongs" in democracy.

In **Zimbabwe**, sacred groves have long been protected by cultural belief. Cutting a tree there isn't illegal — it's unthinkable. These living sanctuaries encode law through reverence, proving that ethics doesn't always need enforcement.

In **Dubai**, designers of *The Green Planet biodome* simulate rainforest conditions to teach urban citizens that nature is not entertainment — it's coexistence. Inside, humidity, soil, and species cohabitate in delicate equilibrium. Each visitor experiences proximity to other lives — the foundation of empathy.

5. Materials That Live and Die Well

If life-centered design begins with biology, it must also consider death.
What happens to materials when their service ends? Can they decompose, feed soil, or seed new forms?

In **Switzerland**, material scientists are creating **circular construction systems** — timber modules that can be disassembled and reused without waste. The principle: nothing permanent, everything continuous.

Dutch designers grow textiles from mycelium; **Scottish innovators** develop seaweed packaging that dissolves in water. **Zimbabwean engineers** make bricks from agricultural waste. Each is an act of humility — acknowledging that nothing we make should outlive its usefulness.

6. The Designer as Steward

The designer of the future is not a creator in isolation, but a **caretaker of interconnection**.

In **Scotland**, landscape architect Julian Jones describes his work as "curating ecosystems."
In **Switzerland**, ETH Zurich's regenerative design courses teach students to "think like a mountain" — a phrase borrowed from Aldo Leopold, meaning to see systems through deep ecological time.
In **Zimbabwe**, young architects design schools that double as rainwater collectors and wildlife corridors — infrastructure as ecosystem.

Stewardship doesn't mean control. It means participation — joining life's complex choreography with awareness and respect.

7. Frameworks for Life-Centered Design

Expansive designers use new tools to embed life in their processes:

- **The Living Systems Map** — charting the species, materials, and cycles a project affects.

- **Ecological Impact Narratives** — telling the story of what the design gives back to nature.
- **Multispecies Charrettes** — workshops where ecologists, farmers, and designers collaborate, speaking for the silent stakeholders.
- **Temporal Testing** — evaluating how a project behaves over seasons, decades, or centuries.

In **Zurich**, architects use *seasonal modeling* to ensure buildings complement migratory birds.
In **Amsterdam**, designers prototype "urban reefs" that regenerate aquatic life.
In **Harare**, permaculture schools train youth to view waste streams as nutrient cycles.
Each method reveals one truth: **life is the client, not the context.**

---

8. Learning from Indigenous and Ancestral Knowledge

Long before "sustainability" became a design term, many cultures already lived by its logic.
In **Zimbabwe**, the Shona and Ndebele traditions honor *chisi* — days when land must rest. No plowing, no cutting, no harvest. The land is allowed to breathe.
In **Scotland**, crofting communities practice shared land management, balancing human need with ecological carrying capacity.
In **Switzerland's alpine pastures**, transhumance — the seasonal movement of livestock — ensures that grasslands regenerate naturally.

These practices are ancient design systems — decentralized, adaptive, reciprocal. The future of design may well depend on remembering them.

9. Life as Measure, Not Metaphor

It's easy to speak of life poetically. Harder to design for it practically.
Yet new metrics are emerging:

- **Net Positive Biodiversity**: a project must increase species abundance where it builds.
- **Ecological Performance Standards**: buildings must behave like native ecosystems.
- **Life Cycle Reciprocity**: materials must return nutrients to soil or industry.

In **Europe**, the *One Planet Living* framework guides cities like Oxfordshire and Geneva.
In **Dubai**, sustainability is becoming law, not luxury.
In **Zimbabwe**, small pilot projects — like eco-brick schools and biogas toilets — prove that life-centered metrics can thrive anywhere, not just in high-tech economies.

10. Closing Reflection: Designing with Humility

When we design for life, we surrender control. We accept that the forest will outgrow our plans, the coral will change color, the wind will sculpt differently than we imagined. And that is good.

Design's greatest evolution is not technological — it's ethical. It is the recognition that *we are not outside of life, but inside it.*

From the **Cairngorms to the Alps**, from **Dubai's desert mangroves to Zimbabwe's forests**, a new lineage of designers is emerging — stewards, healers, ecologists, storytellers. They measure success not by scale or visibility, but by vitality.

To design for all life is to design for **continuity**. For the unseen soil microbes, the returning wolf, the unborn child, and the silent river. It is to say: *We belong to this world — and our design should make that belonging visible.*

Chapter 5 — Dimensions : Designing Beyond Form and Space

"We shape our buildings; thereafter they shape us." — Winston Churchill

The Expanding Field of Design

Design once lived within the confines of geometry — a discipline of lines, forms, and measurable spaces. It was a practice of containment: we drew boundaries, built enclosures, and mastered the manipulation of shape. But the world has changed dimensions.

We now inhabit a time when the physical merges with the digital, when air quality, sound, emotion, and data all shape our experience of place. Form is no longer fixed. A building might breathe, a park might respond, a city might learn. To design beyond form is to design for the invisible — for flows, forces, and feedbacks that transcend three-dimensional space.

1. Beyond Geometry

For much of modern history, design has sought to dominate space. But in the age of systems thinking, we recognize that *space dominates us back*.

In **London**, the *Coal Drops Yard* by Heatherwick Studio turns two Victorian coal warehouses into a marketplace of interaction. The rooflines bend toward each other like hands meeting in midair. It is a space designed not for static function but for encounter — where the geometry itself invites relationship.

In **Scotland**, the *V&A Dundee* by Kengo Kuma channels the rugged cliffs of the Tay Estuary. Its folded concrete walls create shadows that mimic sea caves, making the visitor feel as though they are walking through land and tide simultaneously. Form becomes narrative; structure becomes conversation.

In **Switzerland**, Peter Zumthor's *Therme Vals* submerges visitors in layers of stone and sound. The design's success lies not in its shape but in its rhythm — a slow unfolding of space that mirrors the body's own sense of breath. To design dimensionally is to orchestrate *time, texture, and emotion* as materials.

2. Time as Material

Designers are beginning to treat **time** as a fourth dimension — something to sculpt and stretch.

In **Amsterdam**, floating neighborhoods rise and fall with the tides, blurring permanence and impermanence. In **Zimbabwe**, modular classrooms expand during rainy seasons and contract during droughts.
Across **Europe**, rewilding projects treat patience as design: planting forests that will not mature for a century.

Temporal design resists the tyranny of instant gratification. It invites longevity, aging, and renewal into the creative process. A building that patinates gracefully, a park that shifts with climate, a digital system that evolves with its users — these are the signatures of time-conscious design.

3. Multisensory Worlds

The 20th century trained design to privilege the eye. The 21st century demands that we engage *every sense.*

In **Dubai's Museum of the Future**, corridors hum with light and vibration, immersing visitors in imagined futures through multisensory storytelling. It's not architecture as object but as **organism** — pulsing, breathing, glowing.

In **Zurich**, the *Sound Forest* project transforms environmental data into soundscapes that allow people to "hear" biodiversity. Each tree becomes an instrument; each gust of wind, a musician.

These experiments reveal that the next frontier of design is not visual innovation but **perceptual expansion** — crafting spaces that communicate through touch, resonance, and emotion.

4. Digital and Ecological Dimensions

The line between digital systems and living systems has blurred. Cities now speak in two languages: carbon and code.

In **Scotland**, remote sensors map peatland restoration in real time, visualizing the landscape's breathing cycles. In **Switzerland**, adaptive lighting in public squares adjusts brightness based on moonlight and pedestrian presence.

Even **Dubai's desert gardens** are managed by AI-driven irrigation networks that mimic root systems — watering where plants "signal thirst." Technology, at its best, does not compete with nature but learns from it.

Design beyond form means recognizing that **data is a dimension** of space — a living layer that must be nurtured with the same ethics as soil or stone.

5. Architecture as Ecosystem

Traditional architecture framed the world in edges. Expansive design dissolves them.

In **Berlin**, the *Tegel Urban Tech Republic* repurposes an old airport into a climate-positive district, where rainwater feeds green corridors and buildings act as carbon sinks. In **Zurich**, the *Circle at Zurich Airport* integrates gardens, workplaces, and energy infrastructure into a single breathing complex. The project's central idea: *no dead zones*. Every surface must contribute to life.

Similarly, **Scotland's Eden Project Dundee** (under development) treats buildings as ecosystems — with renewable skins, composting systems, and plant walls that filter air.

The goal is not to build *on* nature but to build *with* it. To see walls, roofs, and foundations as habitats within a greater living web.

6. Emotional Space

There is also the unseen dimension of **feeling** — the psychological geometry of experience. Good design shapes emotion as surely as it shapes space.

In **Harare**, the *Memory House Project* creates courtyards that hold both shade and story — spaces for reflection on ancestral memory.
In **Zurich's hospice gardens**, circular paths lead visitors through sensory landscapes designed to soothe grief. The dimensions of care are measured not in meters, but in moods.

7. Designing in Layers

To design in dimensions is to think in **layers of interaction** — geological, digital, social, and spiritual. Each project becomes a palimpsest, where no layer dominates but all interweave.

In **Dubai**, architects are experimenting with "sand architecture," where desert topography itself determines urban form. In **Switzerland**, train tunnels double as energy conduits, capturing geothermal heat as trains pass. In **Scotland**, artists map wind currents as sculpture, revealing invisible forces that shape the land.

Each layer tells a part of the same story: *design as translation between worlds.*

8. Learning from the Invisible

The most profound dimensions may be the ones we cannot draw — the flows of trust, belonging, memory, and spirit.

Indigenous communities across **Zimbabwe** and **Scotland** design social spaces that mirror ecological ones — circular gathering areas that embody equality, thresholds that honor transition. Their architecture understands what modernism forgot: that the unseen binds the seen.

As we design our future environments — smart, immersive, adaptive — we must also design for the invisible architectures of meaning and connection.

9. Toward Dimensional Ethics

Dimensional design is not only technical; it is ethical. When everything is connected, every form is consequential.

Designers   must   ask:
How   does   this   shape   influence   behavior?
Whose   experience   does   it   privilege?
What unseen systems — ecological or emotional — does it
disturb or heal?

From **Zurich's mindful infrastructure** to **Dubai's responsive
cities**, from **Scotland's landscape architecture** to **Zimbabwe's
vernacular renewal**, the answer is increasingly clear: form is
responsibility.

10. Closing Reflection: The Fifth Dimension

Design beyond form is not about more complexity, but deeper
connection.
The fifth dimension is **awareness** — the understanding that
everything designed participates in a larger choreography of
forces and lives.

Every   curve   alters   air   flow.
Every   material   alters   habitat.
Every choice alters time.

To design dimensionally is to accept that our creations are never
separate from the universe they inhabit — they are *continuations*
of   it.
And in that continuity, the future of design begins to feel less like
invention, and more like revelation.

Chapter 6 — Sectors : Breaking Down Boundaries

"Complex problems do not respect departments." — Anonymous

The Age of Entanglement

Once, design served its own domains. Architects designed buildings, engineers designed bridges, educators designed curricula, and policy-makers designed laws. Each sector defended its turf like a walled city. But climate collapse, inequality, migration, and digital disruption have shown us that the walls leak. Water does not care whether it floods a factory or a school. Carbon does not check if it is entering agriculture or transport.

Expansive design begins where sectors collide — in the messy, fertile edges where disciplines must share roots. To design for all life is to learn the art of **co-creation across boundaries.**

---

1. From Silos to Systems

The twentieth century rewarded specialization; the twenty-first demands synthesis.

In **Scotland**, the *Edinburgh BioQuarter* fuses medicine, urban planning, and landscape ecology. Hospitals open directly onto parks, encouraging patients to recover in daylight and clean air. Public art doubles as environmental sensor. In **Switzerland**, *Circular Economy Switzerland* brings together finance, manufacturing, and waste management to create closed-loop supply chains. Bankers discuss compost. Designers talk about interest rates. Lines blur, solutions emerge.

Across **Europe**, "systems studios" now appear within universities, pairing sociologists with coders, and gardeners with architects. The aim is not efficiency but empathy — learning to see through another discipline's eyes.

2. The Ecology of Collaboration

Collaboration is itself an ecosystem. Each participant plays a role — pollinator, decomposer, seed.

In **Dubai**, the *Desert Farming Initiative* joins marine engineers, climate scientists, and architects to grow crops using seawater greenhouses. The buildings capture humidity by night, condense it by dawn, and feed both plants and people. Agriculture, architecture, and atmosphere co-design.

In **Zimbabwe**, *Solar Villages Network* unites local electricians, craft cooperatives, and teachers. Solar grids power weaving studios by day and classrooms by night. A single infrastructure multiplies purpose.

These projects remind us: innovation happens not inside sectors, but **between** them — in the wild edges where definitions fail.

3. Designing with Policy

Policy is one of the most powerful design tools on Earth, yet it is rarely recognized as such.

Across **Europe**, the *New European Bauhaus* reframes regulation as creative medium. Architects sit beside economists, students beside mayors, prototyping legislation through drawings and models before it becomes law. In **Switzerland**, the *Federal Office for the Environment* invites artists to interpret climate data visually, helping politicians feel what statistics cannot show. In **Scotland**, local councils run participatory budgeting platforms — software that allows citizens to "design" public spending collaboratively. Governance becomes interface.

Design, when merged with policy, gains permanence; policy, when merged with design, gains soul.

4. The Corporate Shift

Even business is re-designing itself as ecosystem.

In **London**, *The Conduit Club* brings social entrepreneurs, scientists, and investors into one circular space where every project must report both profit and planet metrics. In **Zurich**, cooperative start-ups share accounting systems and renewable-energy contracts to reduce collective carbon footprints. In **Dubai**, luxury developers now embed mangrove restoration into real-estate value. The health of the wetland becomes part of the balance sheet.

Cross-sector thinking doesn't dilute purpose — it deepens accountability.

5. Education Without Departments

If the next generation is to think expansively, its classrooms must be designed expansively too.

In **Scotland's Highlands**, secondary students learn climate literacy through fieldwork that combines art, biology, and storytelling.
At **ETH Zurich**, engineering and design students co-develop "living materials" labs with ecologists.
In **Harare**, vocational schools integrate permaculture, architecture, and entrepreneurship, so graduates can build businesses that regenerate soil and society simultaneously.

When education erases its own boundaries, every student becomes a systems designer.

6. The Infrastructure of Connection

Inter-sector design requires infrastructure that encourages mingling — both physical and digital.

In **Dubai's District 2020**, born from Expo 2020, streets are wired for open data sharing between companies, NGOs, and artists.
In **Switzerland**, *Impact Hubs* host scientists, refugees, and financiers at the same tables. Ideas cross-pollinate as naturally as pollen in wind.

Connection is no longer a luxury; it is the new civic utility.

7. Challenges of Collaboration

Boundaries exist for reasons: focus, expertise, identity. Breaking them down demands humility.

Designers must learn to **translate languages** — from spreadsheets to sketches, from poetry to policy. In **Scotland**, environmental designers hold "cross-speak sessions" where every expert must explain their field using a metaphor from nature. In **Zimbabwe**, women's cooperatives use storytelling circles to align agricultural goals with cultural rituals.

The hardest part of collaboration is not disagreement but misunderstanding — and design, as a visual and emotional language, can bridge that gap.

8. Towards Regenerative Economies

The endpoint of cross-sector design is not simply cooperation but regeneration.

In **Europe**, "doughnut-economy" cities like Amsterdam now measure success against ecological ceilings and social foundations. Urban planning, finance, and welfare align under one metric: *thriving within planetary limits.* **Switzerland's Re-Use Valley** recycles industrial waste into modular housing for refugees. Industry and humanitarianism merge seamlessly.

Such efforts redefine prosperity not as growth across sectors, but health across systems.

9. Case Study: The Scottish Blue Network

An inspiring example of sectoral integration is the *Scottish Blue Network* — a collaboration between marine scientists, artists, and fisheries.
Sensors on boats gather data on sea temperature and plastic waste. Artists translate this data into public installations, while policymakers use it to guide coastal zoning. Economy, ecology, and education operate as one design.

The sea becomes not a boundary but a **bridge** between disciplincs.

10. Closing Reflection: The Weave of the World

To design across sectors is to rediscover an ancient truth: the world was never divided in the first place.

Rivers do not differentiate between agriculture and housing. Bees pollinate crops and wildflowers indiscriminately. Human health depends on soil health, which depends on microbial design.

Every attempt to isolate a sector is a simplification of life's complexity.
To design for all life, we must learn to **weave** — to see medicine in architecture, education in agriculture, finance in ecology, and art in governance.

When the weave is whole, design stops being a profession and becomes what it was always meant to be: **a way of keeping the world alive.**

Chapter 7 — Energy : Designing Flows, Not Things

"Energy is the currency of all life." — Buckminster Fuller

The Pulse of the Planet

Everything that lives, moves, or grows depends on energy. But for too long, we have treated it as a commodity rather than a current — something to be extracted, traded, and consumed rather than something that connects, animates, and circulates.

Expansive design asks us to see energy differently: not as power *over* but power *with*. To design for all life is to design for energy as relationship — a choreography of exchange between the human, the ecological, and the technological.

1. From Extraction to Circulation

For centuries, human civilization has drawn energy from the earth as though it were an infinite reservoir. We burned, mined, drilled, and consumed. In doing so, we confused energy with possession — believing that control equaled abundance.

Now, the climate itself reminds us that all extraction has consequence.
Design's new challenge is to create systems where energy circulates — endlessly, gracefully, regeneratively.

In **Switzerland's Monte Rosa Hut**, perched above 2,800 meters, architects designed a self-sufficient building that harvests solar energy, collects meltwater, and manages waste in closed loops. Every watt and drop is recaptured, reused, respected. In **Scotland's Orkney Islands**, community wind farms generate more power than residents can use. The excess is stored as hydrogen — turning rural independence into a renewable export.

Here, design becomes choreography: a dance of sunlight, wind, water, and will.

2. Designing with Nature's Energy

The best energy systems do not imitate machines — they imitate ecosystems.
A forest wastes nothing. A coral reef shares energy across species.
A beehive regulates temperature collectively.

In **Zimbabwe's rural schools**, engineers design *biogas systems* that convert agricultural waste into cooking fuel, providing clean energy while reducing deforestation. Waste becomes warmth; decay becomes nourishment.

In **Dubai**, the *Sustainable City* development powers itself through solar panels integrated into roofs and shades. The layout of streets captures wind and minimizes heat, reducing the need for air conditioning. The result: a modern settlement that breathes like a traditional desert oasis.

And in **Switzerland's Valais region**, hydropower tunnels are being redesigned to mimic natural rivers, creating habitats for fish migration. Even energy infrastructure can serve life rather than sever it.

3. Social Energy

Not all energy is electrical. Some of it is emotional — the spark that flows between people when they create, share, or rebuild.

In **Harare's women's cooperatives**, communal kitchens powered by solar stoves become hubs of learning and laughter. Energy fuels not only food but friendship. In **London**, community centres retrofit old buildings with heat pumps and rooftop gardens, using design as an invitation for neighbours to work side by side. In **Scotland**, energy cooperatives invest profits into local childcare and elder care — converting wind power into social warmth.

Designers who recognize social energy understand that the grid is not just wires; it's relationships.

4. Cities as Power Plants

The cities of the future will not just consume energy — they will generate it.

In **Zurich**, district heating networks reclaim waste heat from data centers to warm residential buildings. In **Amsterdam**, floating solar panels move with the tide. In **Dubai**, kinetic pavements turn footsteps into electricity, while shaded walkways reduce heat absorption.

Every surface, from a window to a wall, becomes a potential generator.
Every design decision becomes an act of stewardship — a way to feed, not deplete, the shared grid of life.

5. Invisible Energy: Data and Flow

Energy now moves through invisible channels as much as visible ones.
Data, too, is a form of power — shaping behaviour, economies, and governance.

But data centres consume vast amounts of electricity. Designers in **Switzerland** are now creating "symbiotic servers" — computer farms that heat nearby homes through distributed networks.
In **Dubai**, algorithms balance solar output and human usage across neighbourhoods, ensuring equitable distribution.
And in **Scotland**, marine turbines double as ocean observatories, transmitting ecological data along with electrical current.

When data and energy merge responsibly, they can form **a nervous system for sustainability** — sensing, learning, and adjusting in real time.

6. Designing Rhythms, Not Machines

Energy is not a static resource — it is rhythm. Daylight shifts, tides rise, seasons breathe.
Designers are beginning to work with these natural cadences instead of against them.

In **Zurich**, office towers dim automatically at sunset and brighten with dawn, aligning human circadian cycles with solar rhythm.
In **Scotland**, rural homes use phase-change materials that store heat during the day and release it at night.

In **Zimbabwe**, farmers' irrigation systems are powered by slow-drip solar pumps that match water flow to sunlight intensity — a literal harmony of elements.

Such designs remind us that efficiency is not acceleration — it is alignment.

7. The Ethics of Energy

Every energy system expresses a worldview. Fossil fuels said: "The earth is ours to burn." Renewables say: "The earth is ours to balance."

But balance requires justice. In **Africa**, millions still live without reliable electricity, while excess energy is wasted in global capitals. Life-centered design asks: *Who benefits from the light?*

In **Zimbabwe**, local energy cooperatives train youth as solar technicians, decentralizing expertise and income. In **Scotland**, community wind farms reinvest profit locally instead of feeding distant shareholders. In **Switzerland**, cities debate "energy citizenship" — giving every resident a legal right to generate, store, and share renewable power.

Energy equity is not a technical problem; it is a design problem.

8. Energy Literacy and Beauty

To design for all life, we must make energy visible, tangible, and beautiful.

In **Dubai's Expo Pavilion "Terra"**, massive solar "trees" rotate with the sun, making sustainability an experience of wonder rather than guilt.
In **Zurich's ETH Campus**, transparent pipes carry recycled water along glass walls, turning infrastructure into education.
In **Scotland**, community art projects use glowing installations powered by bike pedaling — a reminder that motion and joy can coexist with conservation.

When people can see and feel energy, they begin to care for it.

9. Regenerative Flow

Nature runs on loops, not lines.
Designers inspired by this principle are now creating *regenerative systems* that restore more energy than they consume.

In **Switzerland**, experimental algae panels produce biofuel while absorbing carbon dioxide.
In **Zimbabwe**, small hydropower installations are built alongside fish ladders and wetland restoration, proving technology can heal ecosystems rather than harm them.
And in **Dubai**, desert buildings use solar chimneys to cool interiors passively, reducing reliance on mechanical systems entirely.

Regeneration is not about doing less harm; it is about doing more good.

10. Closing Reflection: The Flow of Life

To design for energy is to design for flow. From sunlight through leaf to breath, from tide through turbine to lamp, from effort through cooperation to joy — energy connects all that exists.

When we treat energy as a living cycle, not a lifeless commodity, our designs begin to mimic the elegance of the natural world. Buildings become lungs, streets become arteries, and communities become beating hearts.

The true measure of progress is not how much energy we consume, but how gracefully it moves through us. And in that grace lies the blueprint for a planet that thrives — endlessly, beautifully, and alive.

Chapter 8 — Culture : Designing with Stories and Identities

"Every design is a story about who we think we are." — Anne-Marie Willis

Design as Storytelling

Culture is design's invisible foundation. Every pattern, pathway, and product carries a story — about belonging, memory, power, and   purpose.
When designers ignore culture, they erase context. When they embrace it, design becomes a living narrative — one that connects the old and the new, the local and the global, the human and the more-than-human.

Expansive design understands culture not as decoration, but as DNA.
It asks: *Whose stories are being told?* And *whose stories are still waiting to be heard?*

1. The Material of Memory

In **Scotland**, the *St Kilda Centre* on the Isle of Harris honors a vanished island community through design that echoes the rhythms of sea and stone. Every wall seems to listen to the wind. Materials are sourced from nearby cliffs, keeping memory close to   matter.
In **Zimbabwe's Matobo Hills**, granite shelters hold rock art created thousands of years ago — ancient design that still speaks through pigment and form. Local architects now draw inspiration from these enclosures to create community halls that feel both contemporary and ancestral.

Design grounded in memory reminds us that progress is not a race away from the past, but a return to the stories that shaped us.

2. Living Heritage

Culture is not static; it is continuously redesigned. In **Switzerland's Alpine valleys**, architects build chalets using ancient wood-joint techniques combined with modern insulation — preserving the integrity of craft while meeting contemporary needs.
In **Dubai's Al Fahidi District**, courtyards are shaded by mashrabiya screens carved using computer-guided tools. Technology revives tradition, translating the aesthetic of shadow into a digital craft language.

Across **Europe**, artisans collaborate with designers to update endangered skills: lace weaving in Bruges becomes 3D-printed textile; Scottish tweed is now woven with recycled ocean plastics.

Culture survives not by freezing itself, but by **flowing forward**.

3. Design as Cultural Repair

Design can also heal when culture has been fractured.

In **Zimbabwe**, colonialism separated people from land and language. Community projects like *Njelele Art Station* in Harare are reclaiming public space as storytelling space — murals, performances, and architecture that reweave identity with landscape.
In **Bosnia and Rwanda**, memorials are designed with community

participation, blending architecture and ritual. Here, design becomes a form of mourning and renewal.

And in **Scotland**, Gaelic language schools use architecture to embody linguistic revival: circular classrooms encourage oral storytelling, echoing the ceòl mòr (great music) of Gaelic song.

Culture-cantered design does not just make objects — it makes *wholeness*.

4. Cross-Cultural Collaboration

In a globalized world, designers are cultural translators.

In **Zurich**, the *Design Biennale* pairs Swiss designers with African and Middle Eastern makers to co-create works around climate adaptation. The results are hybrid artifacts — half science, half myth.
In **Dubai's Design District (d3)**, calligraphers, coders, and fashion designers share open studios, fusing Arabic script with augmented reality. Tradition becomes a launchpad for innovation. Meanwhile, **Zimbabwean diaspora designers in London** are weaving African pattern systems into high-tech sportswear — a fusion of rhythm and resilience.

Cross-cultural design dissolves hierarchy; it transforms culture into dialogue.

5. Architecture as Language

Buildings are the most public stories a culture tells. In **Scotland**, the *V&A Dundee* reimagines the country's coastal cliffs as cultural memory cast in concrete. In **Switzerland**, Peter Zumthor's chapels whisper of silence and stone, embodying humility as aesthetic. In **Dubai**, mosques with solar panels now merge spirituality with sustainability — a quiet conversation between the sacred and the scientific.

Each project becomes an accent in the world's architectural language, speaking of what its people value most.

6. Design Against Cultural Erasure

As the world urbanizes, cultures risk being flattened into sameness.
Life-centered design resists this monoculture by designing for plurality.

In **Zimbabwe's rural markets**, local crafts are protected through digital mapping that links artisans directly to global buyers — preventing cultural knowledge from disappearing. In **Scotland**, heritage groups use VR to preserve endangered dialects and folk rituals. In **Switzerland**, museums now co-curate exhibitions with immigrant communities, expanding the definition of "Swissness" beyond the Alps.

Design here becomes activism — a defense of diversity as life itself.

7. Ritual and the Everyday

Culture is not only in museums or monuments. It lives in gestures: how we gather, eat, greet, and celebrate.

In **Dubai**, contemporary designers reinterpret the majlis — the traditional gathering space — as a circular lounge open to all genders and generations.
In **Scotland**, community bakeries design bread ovens that serve as social hearths.
In **Zimbabwe**, water pumps are painted in bright geometric patterns by local children, turning infrastructure into play.

When everyday rituals are honoured, design rekindles belonging.

8. Learning from Indigenous Design

Indigenous design is the original life-centered practice. It holds centuries of knowledge about sustainability, reciprocity, and place.

In **Zimbabwe**, the Shona concept of *hunhu* — the belief that "a person is a person through other people" — inspires architecture that prioritizes communal over individual use.
In **Scotland**, crofting communities share land through collective tenure, designing stewardship into law.
In **Switzerland's alpine villages**, seasonal migrations of cattle between valleys (transhumance) maintain soil health and social rhythm.

These traditions are not nostalgia — they are **design systems tested by time**.
Their logic of balance offers a compass for the future.

9. Culture as Catalyst for Sustainability

Design rooted in culture naturally sustains.
In **Europe's New European Bauhaus** initiative, beauty and sustainability are seen as inseparable — a cultural shift from utility to meaning.
In **Dubai**, fashion brands adopt desert tones and organic materials, linking aesthetics to ecology.
In **Harare**, musicians collaborate with architects to design sound-based installations powered by solar energy — blending rhythm, art, and activism.

Culture makes sustainability desirable because it makes it *feel like us*.

10. Closing Reflection: Stories that Keep the World Alive

Culture is not a layer we add to design — it is the soil from which design grows.
When we design with stories, we design with soul.

From **the granite echoes of Matobo** to **the luminous towers of Dubai**, from **the Gaelic rhythms of the Hebrides** to **the alpine whispers of Switzerland**, we see that every place carries wisdom worth protecting.

To design for all life is to honor those stories — to let materials speak memory, to let form express empathy, to let spaces remember what we too easily forget:
That culture is not what divides us.

It is what connects us — the shared pulse of imagination that keeps the living world alive.

Chapter 9 — Intelligence : Designing with Human, Artificial, and Collective Minds

"Intelligence is not something we possess, but something we participate in." — David Abram

The Expanded Mind

Design is a conversation between intelligences — human, natural, and now artificial. The question is no longer *whether* machines can think, but *how* we think together. In nature, intelligence is everywhere: in the murmuration of starlings, the mycelial networks beneath forests, the migration patterns of whales, the precision of bees. These are not metaphors — they are real forms of knowing.

Expansive design asks us to recognize intelligence as distributed, relational, and ecological. It does not reside in a single brain or algorithm but in the *connections between things*.

To design for all life is to design with all minds.

1. From Artificial Intelligence to Ecological Intelligence

Today, artificial intelligence shapes everything from urban planning to material research. But intelligence without ecology risks becoming clever without wisdom.

In **Switzerland**, the *ETH Zurich Digital Ecology Lab* is pioneering *Ecological AI* — algorithms that learn from natural systems, modeling the self-balancing dynamics of forests to optimize building energy flows. The aim: not domination, but *symbiosis*.

In **Dubai**, AI monitors desert flora to predict microclimate shifts and guide urban planting — using computation as a listening device rather than a control system.

And in **Zimbabwe**, young coders at *Muzinda Hub* are building AI tools for small-scale farmers to predict rainfall and pest outbreaks, blending digital intelligence with ancestral knowledge of the land.

Intelligence becomes expansive when it aligns with the rhythms of life rather than replacing them.

2. The Collective Mind

The greatest designs are not born from solitary genius but from collective intelligence.

In **Scotland's Highlands**, community-driven mapping projects invite residents to record ecological memories — where salmon once ran, where peat once healed. These living maps inform restoration design far more accurately than satellite data.

In **Switzerland**, *co-labs* bring together biologists, data scientists, and artists to design climate-responsive materials. In **Zimbabwe's Mutare region**, village councils use participatory design platforms — digital tools translated into Shona — to co-create infrastructure plans.

Here, design is not something *done for* people but *done with* them. Collective intelligence transforms design from authorship into stewardship.

3. Intelligence Beyond the Human

For centuries, intelligence was defined by language, logic, and technology — all human measures. But forests sense, corals communicate, and slime molds solve mazes.

In **Switzerland's Lausanne University**, researchers are studying plant signaling as a model for decentralized design: each leaf responds autonomously, yet the whole tree maintains coherence.

In **Zimbabwe's wetlands**, ecologists design irrigation channels based on termite mound logic — networks that distribute moisture evenly, maintaining stability in drought.

Nature is not just inspiration — it is **instruction**. Designers who listen to it learn that intelligence can be slow, silent, and embodied.

4. Ethical Intelligence

Design guided by intelligence must also be guided by ethics.

In **Dubai**, the rise of AI-driven architecture raises questions about authorship, labour, and privacy. Who owns the designs created by machines?
In **Europe**, the *AI Act* establishes ethical frameworks for transparency and accountability — recognizing that the smartest

systems can still cause harm if they amplify bias. And in **Zimbabwe**, where facial recognition is being trialed in urban security, activists advocate for "Ubuntu AI" — technology built on communal consent and care.

Ethical intelligence means remembering that not all that can be done should be done. Wisdom is the intelligence of restraint.

5. Learning from Collective Systems

Intelligence thrives in feedback loops. In **Switzerland**, energy grids now use AI to balance supply and demand minute by minute, learning from usage patterns. In **Scotland**, oceanic data from buoys, fishers, and satellites feed into shared platforms for marine conservation — collective awareness turned into adaptive policy. In **Dubai**, smart irrigation systems use AI to mimic natural water cycles, releasing moisture based on soil feedback.

These systems reveal a truth that nature has always known: intelligence is not control — it is relationship.

6. The Role of the Designer in an Age of AI

When AI can generate images, structures, and strategies in seconds, what is left for human designers to do? Everything that matters.

The designer of the future will not compete with machines but **curate** their intelligence — framing questions, defining ethics, and embedding empathy.

In **Switzerland**, architectural students collaborate with generative AI to simulate centuries of climate impact — using machines not as replacements but as foresight partners.
In **Scotland**, designers use AI to reconstruct lost Gaelic sounds for cultural installations — reviving language through technology.
In **Zimbabwe**, artists use neural networks to remix oral histories into digital tapestries, teaching AI to speak with, not over, the voices of ancestors.

The new designer is not a maker, but a *mediator* — weaving human and non-human minds into coherence.

7. Slow Intelligence

Not all intelligence is fast. Some of the wisest systems evolve over millennia.
The glacier, the forest, the coral reef — each thinks in time spans longer than empires.

In **Switzerland**, scientists studying glacier retreat are partnering with philosophers to redefine intelligence as "the capacity to maintain equilibrium."
In **Scotland's Flow Country peatlands**, researchers map moss growth over decades, understanding slowness as resilience.
In **Zimbabwe's baobab forests**, communities read the trees as calendars, weather stations, and teachers.

Designers who work with time as a collaborator begin to think like ecosystems — patient, adaptive, alive.

8. Data as Ecology

Data is often imagined as neutral. But it, too, has an ecology —
an extraction footprint, an energy cost, and a moral weight.

In **Zurich**, sustainable AI labs are designing low-energy models
trained on smaller, local datasets to reduce computational
emissions.
In **Dubai**, architects visualize real-time air quality data as shifting
light sculptures, turning invisible information into public
awareness.
In **Zimbabwe**, civic tech groups use open data to track
deforestation and empower local action.

When data becomes dialogue rather than domination, intelligence
becomes ecological again.

9. Coexisting Intelligences

The challenge ahead is not to prove that artificial or natural
intelligence is superior, but to design systems where they coexist.

Imagine buildings that learn from weather patterns like trees.
Imagine city grids that communicate like coral reefs.
Imagine economies that adapt like bee colonies.

This is not fantasy — it is already beginning.
In **Switzerland**, AI-powered transport networks respond
dynamically to wildlife movement data, reducing collisions.
In **Scotland**, forest management uses sensor networks that alert
rangers to drought stress in real time.

In **Zimbabwe**, solar microgrids autonomously balance community energy needs using simple neural logic.

These are glimpses of a world where human, artificial, and biological intelligence are woven together in mutual awareness.

10. Closing Reflection: The Mind of the Living World

The future of intelligence is not artificial — it is *expanded.* Every living system already knows how to self-organize, adapt, and evolve. The task of design is to align with that wisdom, not outpace it.

As we code algorithms, model materials, and build systems, we must remember that the smartest designs are not those that think faster — but those that think **with**.

The mountain knows gravity.
The bird knows geometry.
The river knows time.

When we design in conversation with these intelligences — human, machine, and more-than-human — we begin to glimpse the mind of the living world itself. And that, perhaps, is the highest form of design: Not to outthink life, but to think *as* life.

Chapter 10 — Ethics : Designing with Responsibility, Equity, and Reverence

"The first rule of sustainability is to align with natural forces, or at least not try to defy them." — Paul Hawken

The Moral Geometry of Design

Every design is a moral act. Even a chair expresses ethics: who it's made for, who made it, what resources it consumed, and what it leaves behind. To design for all life is to make ethics not an afterthought, but the foundation.

Expansive ethics recognizes that design does not merely shape form — it shapes futures. It asks: **Who benefits? Who bears the cost? And what kind of world does this create when we are gone?**

Ethics, in the age of planetary design, is no longer about compliance — it is about *conscience.*

1. From Sustainability to Responsibility

For decades, "sustainability" has been design's moral shorthand. Yet sustaining a damaged system is not enough. The new ethic is **responsibility** — the ability to respond wisely, humbly, and justly.

In **Switzerland**, architects of the *ReSource Pavilion* use reclaimed timber and steel, publicly labeling every component's past life. Each piece carries a story of material resurrection — a

testament to accountability.
In **Scotland**, community-led housing trusts prioritize local materials and fair labor over profit, designing dwellings that build social resilience as much as shelter.
In **Zimbabwe**, grassroots builders craft structures from earth and stone, balancing modern safety codes with ancestral methods that require no imported cement.

Responsibility here means designing in dialogue with time — honouring both inheritance and impact.

2. Equity as Design Principle

An ethical design does not merely sustain ecosystems; it sustains *justice*.
The climate crisis is not evenly distributed, nor should design's benefits be.

In **Dubai**, labour reforms in construction are slowly embedding fairness into the world's most ambitious skylines. Ethical design means reimagining the worker's place in the process — not as invisible labour, but as co-creator.
In **Switzerland**, urban renewal in Zurich's Kreis 5 ensures affordable housing amidst gentrification through cooperative ownership models.
And in **Zimbabwe's rural electrification projects**, solar cooperatives train local women as technicians, placing energy and agency in the same hands.

Design becomes equitable when it shares power as generously as it shares beauty.

3. Transparency and Traceability

Design ethics must be visible — literally.

In **Europe**, the *Materials Passport* initiative tracks the origin, carbon cost, and recyclability of every construction component. Buildings now carry "ethical blueprints." In **Switzerland**, furniture labels display water usage and biodiversity impact alongside price tags. In **Zimbabwe**, mobile apps connect farmers directly to consumers, eliminating exploitative middlemen and revealing every step of production.

Transparency turns design into storytelling — not just of what something *is*, but of what it *means*.

4. The Right to Repair and Regenerate

Ethics is also about longevity — resisting the culture of disposability that defines modern consumption.

In **Scotland**, "repair cafés" let citizens fix electronics and furniture collectively. Designers attend these sessions, learning from failure and from users' ingenuity. In **Switzerland**, fashion brands like Freitag and Qwstion design bags to be repaired for life, not replaced. In **Zimbabwe**, tailors remake old uniforms into schoolbags, turning waste into dignity.

Ethical design reclaims time itself — slowing the rhythm of making to match the rhythm of living.

5. Designing for the Unseen

The most powerful ethical shift happens when we extend empathy beyond the visible.

In **Dubai**, desert architects plan with wind and shade in mind, reducing energy while honoring the landscape's fragility.
In **Scotland**, offshore wind farms are mapped with migratory bird routes to minimize ecological disruption.
In **Zimbabwe**, borehole projects include animal watering stations, ensuring that wildlife benefits from human design.

When ethics includes the non-human, design becomes a prayer — an offering to the unseen systems that sustain us.

6. Digital Ethics: The Invisible Frontier

As design merges with data, new moral questions emerge. Who owns the data of a smart city? How transparent are the algorithms shaping our environments?

In **Zurich**, designers advocate for "data dignity" — citizens owning the data generated by their movements.
In **Dubai**, smart city dashboards are now being re-evaluated for accessibility and consent.
In **Zimbabwe**, civic hackers create open-source tools that map pollution and corruption, proving that digital ethics can also be resistance.

Ethical design in the digital realm demands one guiding principle: technology should amplify human and ecological dignity, never diminish it.

7. Designing Through Care

Ethics is not only a framework — it is a feeling. Care is the emotional intelligence of design.

In **Scotland**, care homes are co-designed with residents, ensuring that architecture feels like a companion rather than an institution. In **Switzerland**, hospitals integrate gardens and daylight to support healing through biophilic design. In **Zimbabwe**, mothers' shelters near maternity wards are shaped with open courtyards and shaded communal spaces — architecture as empathy.

To design with care is to treat every life as sacred, every need as narrative.

8. Decolonizing Design Ethics

Ethics must also confront history. Modern design owes much of its material wealth to colonial extraction and inequity. To design ethically is to reckon with that inheritance.

In **Zimbabwe**, cultural centres like *Nhimbe Trust* use architecture to preserve indigenous storytelling and ritual — not as nostalgia, but as a form of sovereignty. In **Europe**, museums are returning looted artifacts, redesigning their own role from possession to partnership. In **Switzerland**, educational institutions are re-examining their colonial-era collections, embedding restitution into pedagogy.

Decolonizing design is not about guilt — it is about gratitude and repair.

9. Reverence: The Forgotten Ethic

Beyond justice lies reverence — a sense of humility before the complexity   of   life.
Design guided by reverence knows when not to build, when to leave a landscape untouched.

In **Scotland's Isle of Eigg**, the community rejected overdevelopment, opting for small, renewable-powered dwellings   that   protect   the   island's   spirit.
In **Switzerland**, mountain architects refuse commissions that would   scar   fragile   ecosystems.
In **Zimbabwe's sacred forests**, villagers still perform rituals before harvesting wood — acknowledging the exchange between life and use.

Reverence turns ethics into art — the art of knowing our place in the living web.

10. Closing Reflection: The Conscience of Design

Ethics is not a constraint on creativity — it is its highest form. It transforms design from the making of things into the **care of life**.

From **the wind farms of Scotland** to **the markets of Harare**, from **the alpine pavilions of Zurich** to **the desert sanctuaries

of Dubai, a new generation of designers is discovering that morality and beauty are not opposites but twins.

To design ethically is to love wisely. It is to remember that every blueprint is a promise — to the present, to the future, and to all that breathes between them.

The measure of design, in the end, is not how much it dazzles, but how deeply it honors. And when ethics becomes instinct, design becomes devotion.

Chapter 11 — Futures : Designing for the Next Thousand Years

"We do not inherit the Earth from our ancestors; we borrow it from our children." — Native American proverb

Design as Time Travel

Design is always an act of time travel. Every line we draw, every structure we build, every product we create is a message to the future — saying, *This is who we were. This is what we valued. This is what we left behind.*

Most design today thinks in product cycles or political terms — five years, ten years, a fiscal quarter. But to design for all life, we must think in **planetary time** — centuries, millennia, the slow unfolding of ecosystems.

The question is no longer *What will this design achieve?* It is *What will it sustain — and who will it serve — long after we are gone?*

1. Deep Time Thinking

In **Switzerland**, the *Gotthard Base Tunnel* was designed to last a century — a feat of precision and endurance. But what if our designs could last a thousand years, not through strength alone, but through adaptability?

In **Scotland**, architects studying Neolithic stone circles are rediscovering how early builders designed with celestial cycles, not calendars. They built in relationship to the sun, moon, and solstice — not as superstition, but as timekeeping in stone.

In **Zimbabwe**, the Great Zimbabwe ruins remind us that architecture can hold continuity even in decay. The stones breathe with history; the form persists through erosion.

Designing for deep time means accepting change as material. It is to build for transformation, not permanence.

2. Futures as Multiverse

There is no single future — only **futures**: some near, some distant, some hopeful, some fragile.

In **Dubai**, designers of the *Museum of the Future* imagined a world of sustainable progress, where technology and nature coexist.
In **Switzerland**, foresight labs map scenarios for 2100 — from rewilded Alps to energy-neutral cities.
In **Scotland**, artists work with climate scientists to visualize what their coasts might look like in 2300 — a merging of imagination and responsibility.

And in **Zimbabwe**, futurists are reimagining African progress not as imitation of the West, but as a leapfrog evolution — cities powered by sun and soil, rooted in *Ubuntu*, guided by community.

Design for all life embraces plural futures — not to predict them, but to prepare for them.

3. Designing for Regeneration, Not Growth

The dominant story of design has long been growth — bigger, faster, more.
But the next thousand years demand a new story: **regeneration**.

In **Europe**, circular economies are becoming policy — mandating that what is made must one day be unmade and reborn.
In **Switzerland**, regenerative agriculture and architecture are converging — buildings that compost, farms that build soil faster than they deplete it.
In **Zimbabwe**, permaculture villages turn waste into wealth, food into forest.
And in **Scotland**, rewilding estates restore wolves, trees, and rivers to their natural intelligence.

Design for regeneration aligns human creativity with nature's metabolism — design as compost, not conquest.

4. Designing for Uncertainty

The future is uncertain — and that's its beauty.
Design cannot control what comes next, but it can create conditions for resilience.

In **Zurich**, engineers design flood-adaptive infrastructure — bridges that float rather than resist.
In **Dubai**, modular desert dwellings can expand or contract with temperature and population.
In **Scotland**, homes in coastal areas are built to move — to be relocated as the sea reclaims its space.
In **Zimbabwe**, drought-resistant community gardens use local seeds that have evolved resilience over centuries.

Uncertainty invites flexibility. The most enduring designs are those that can change gracefully.

5. The Ethics of Legacy

What we make today becomes someone's inheritance tomorrow. Legacy is not measured in monuments but in continuance — the quiet persistence of systems that nurture life.

In **Switzerland**, glacier memorials mark the places where ice once was, teaching future generations to remember what vanished.
In **Scotland**, young designers plant "design forests" — each tree representing a project meant to grow alongside the planet, not at its   expense.
In **Zimbabwe**, architects design schools that double as seed banks — places where education and ecology intertwine.
And in **Dubai**, innovators experiment with desert architecture that produces more water than it consumes — abundance as legacy.

Legacy is the morality of time. It asks not, "What can I build?" but "What can I leave living?"

6. Rewilding Imagination

Before we can rewild the planet, we must rewild our imagination. Industrial modernity taught us to see nature as background. The future of design will teach us to see it as collaborator, kin, and teacher.

In **Europe**, designers are reintroducing myth — dragons as metaphors for data, forests as archives of wisdom.
In **Zimbabwe**, oral storytelling revives ecological literacy — parables that encode how to live with rain, wind, and drought.
In **Scotland**, speculative design studios invite children to imagine "Gaia 3025" — not as apocalypse, but as a thriving, multispecies civilization.

Rewilded imagination doesn't fear the unknown — it learns to dream with it.

7. Designing with the Future Present

Time is not linear; it is ecological — cycles within cycles, each influencing the other.
To design with the future in mind is to treat it as already present.

In **Switzerland**, designers use climate models to shape today's architecture as though they were building for 2080.
In **Dubai**, vertical gardens simulate future heatwaves to test plant survival.
In **Zimbabwe**, architects design for future generations by building with youth — letting children place the first stone.

Every future is built in the now. Each ethical choice, each material restraint, each gesture of reverence is a seed planted in tomorrow's soil.

8. Global Responsibility, Local Expression

The design challenges of the future — climate, biodiversity, equity — are global. But their solutions will always be local.

In **Switzerland**, alpine villages redesign tourism to honor silence and slowness.
In **Scotland**, islanders craft their own renewable grids, proving sovereignty through self-sufficiency.
In **Dubai**, water-scarce neighborhoods develop circular reuse systems rooted in Bedouin wisdom.
In **Zimbabwe**, indigenous councils manage land through consensus, blending technology with tradition.

Future design will succeed not through uniformity, but through **diversity in coherence** — the planet thinking in many dialects, yet one language of care.

9. The Thousand-Year Designer

To design for the next thousand years is to design with humility. The thousand-year designer plants trees whose shade they will never sit in, builds schools whose graduates they will never meet, and restores lands whose full flourishing they will never see.

In **Switzerland**, long-term material research projects track wood aging over decades.
In **Scotland**, intergenerational design councils include children as decision-makers for public works.
In **Zimbabwe**, community archives record not just structures, but values — a blueprint for spirit as much as space.

The thousand-year designer is not an architect of monuments but of continuity.

10. Closing Reflection: The Future Is Alive

The future is not a destination — it is a relationship. Every decision we make now sends ripples across centuries, shaping the conditions for life yet unborn.

From **the highlands of Scotland** to **the Alps of Switzerland**, from **the deserts of Dubai** to **the savannas of Zimbabwe**, we are learning that the future is not something we enter — it's something we cultivate.

Designing for all life means designing for the living continuum — the soil that remembers, the rivers that learn, the cities that grow kinder with age.

Let our blueprints breathe.
Let our materials return to the earth gracefully.
Let our ethics outlast our egos.

If we can do that, then a thousand years from now — when someone touches a wall we built, walks a path we restored, or drinks from a well we once designed — they will not remember our names.
They will feel our intention.

And perhaps they will say, quietly, "They designed as if the world were alive — and so it remained."